HUNGARY

MAJOR WORLD NATIONS

HUNGARY

Julian Popescu

CHELSEA HOUSE PUBLISHERS
Philadelphia

Chelsea House Publishers

Printed in Malaysia

First Printing.

1 3 5 7 9 8 6 4 2

Library of Congress Cataloging-in-Publication Data

Popescu, Julian.
Hungary / Julian, Popescu
p. cm. — (Major world nations)
Includes index.
Summary: An overview of the history, geography, economy, government, people, and culture of Hungary
ISBN 0-7910-5386-5 (hc.)
1. Hungary—Juvenile literature. [1. Hungary.] I. Title.
II. Series.
DB906.P67 1999
943.9—dc21 99-19153
CIP

ACKNOWLEDGEMENTS

The Author and Publishers are grateful to the following organizations and individuals for permission to reproduce the illustrations in this book:
Colorpix; Mike Dixon; The Hungarian Embassy; INTERFOTO MTI, Hungary; The Mansell Collection Ltd; Novosti Press Agency; D. C. Williamson - London.

CONTENTS

		Page
Map		6
Facts at a Glance		7
History at a Glance		9
Chapter 1	In Central Europe	13
Chapter 2	Land and Climate	19
Chapter 3	On the Banks of the Danube	25
Chapter 4	Early History and Kings	31
Chapter 5	The Austro-Hungarian Empire	39
Chapter 6	Recent History	45
Chapter 7	Industry and Transportation	51
Chapter 8	Forests and Fisheries	56
Chapter 9	Farms and Crops	59
Chapter 10	*Csikos* and the Horse Round-up	67
Chapter 11	Budapest and Other Cities	71
Chapter 12	Schools and Sports	81
Chapter 13	Artists and Scientists	85
Chapter 14	Rights and Duties	90
Chapter 15	Hungary and the Modern World	94
Glossary		98
Index		100

HUNGARY
POLAND
SLOVAKIA
UKRA
USTRIA
NORTHERN MTS.
Tokaj
Miskolc
AGGTELEK MTS.
Eger
L.Fertö
Györ
R. Tisza
Debrecen
BUDAPEST
LITTLE PLAINS
R. Raba
BAKONY FOREST
Szolnok
Zalaegerszeg
R. Danube
ALFÖLD
L.Balaton
TRANSYLVA
MECSEK MTS.
Szeged
R. Maros
ROMANIA
Pécs
SLOVENIA
R. Dráva
CROATIA
SERBIA
N
0 50 100 150 km
0 50 100 miles

FACTS AT A GLANCE

Land and People

Official Name	Republic of Hungary
Location	Central Europe
Area	36,000 square miles (92, 103 square kilometers)
Climate	Temperate
Capital	Budapest
Other Cities	Miskolc, Debrecen, Szeged, Pécs
Population	10, 208, 000 (1998)
Major Rivers	Danube
Major Lakes	Balaton, Fertö
Mountains	Carpathian, Mátra, Bükk, Mecsek, Aggtelek
Highest Point	Mount Kékes (3,350 feet / 1,015 meters)
Official Language	Hungarian
Ethnic Groups	Magyars (Hungarians), 95 percent; Gypsies

Religions	Roman Catholic, 67 percent; Reformed (Calvinist), 20 percent
Literacy Rate	99 percent
Average Life Expectancy	66.46 years (male); 75.44 years (female)

Economy

Natural Resources	Bauxite, coal
Division of Labor Force	Services, 65 percent; industry, 26 percent; Agriculture, 8.3 percent
Agricultural Products	Grain, meat, rice, wheat, sugar beet, potatoes
Industries	Manufacturing, chemicals, mining
Major Imports	Oil, natural gas, agricultural products
Major Exports	Meat, agricultural products, pharmaceuticals, bauxite
Currency	Forint

Government

Form of Government	Republic
Government Bodies	National Assembly (parliament); Council of Ministers; Supreme Court
Formal Head of State	President
Head of Government	Prime Minister
Other Chief Officials	Ministers
Voting Rights	All persons 18 years of age or older

HISTORY AT A GLANCE

3rd century B.C. The Romans conquer the lands surrounding the Danube River calling the area the province of Pannonia (present day Transdanubia, Hungary).

1st century B.C. The Romans abandon their holdings in the Danube area due to repeated attacks by barbarians.

5th century A.D. Attila the Hun and his men roam the plains of Europe destroying villages and killing their inhabitants.

9th century The Great Migration begins with the Magyars (Hungarians) leaving their homeland in the north and eventually traveling over the Carpathian mountains and settling on the plains. They proceed to raid their neighbors in Bulgaria, Italy, Germany, and Poland and are much feared for their warlike temperament and archery skills.

955 The German knights fight the Hungarians at the battle of Lechfeld, finally defeating them.

997-1083 King Stephen I gains the throne of Hungary and unites all the Magyar tribes into one nation. King Stephen also converts his people to Christianity, for which he later is declared a saint by the pope.

1241 Hungary is invaded by the Mongols, who kill and plunder its people and villages.

1415 The Hungarian preacher Jan Huss is brought to trial by the Roman Catholic Church and found to be a heretic. He is then burned at the stake.

1526-1547 The battle of Mohacs is fought in 1526 against the Turks. The Turks are victorious and continue their invasion of Hungary, conquering all of the southern and central area. The Turks and their Ottoman Empire would stay for over 150 years in Hungary.

1569 The Hapsburg Maximillian II signs a treaty with the Turks in which Hungary is divided into three parts. The north is made the kingdom of Hungary; in the east Transylvania is made a principality; central and southern Hungary remain under the rule of the Turks.

1683-1699 The Polish army aids the Hungarians in their battle against the Turks who have been trying to acquire more of Hungary's lands.

1699 A treaty is signed at Karlowitz in Serbia in which the Turkish sultan gives up his claim to Hungary and makes Hungary a province of the Austrian Hapsburg empire.

1740 Empress Maria Theresa begins her 40 year reign of the Hapsburg Empire. She levies many taxes against the Hungarians causing widespread discontent.

1832-1836 Lajos Kossuth wages a campaign demanding reforms for the Hungarian people. He is eventually imprisoned.

1848 Kossuth leads an all-out revolt against the Austrian Empire but the war for independence is lost.

1856 European powers sign a treaty stating that the Danube River is a free highway to the ships of all countries, and form an International Commission to regulate its navigation, settle any disputes, and maintain the river.

1867 Austrian emperor Francis Joseph is forced to recognize Hungary as a self-governing kingdom and the combined countries are now called the Austro-Hungarian Empire. A Hungarian government takes care of domestic affairs while the Austrian monarch handles foreign affairs and defense.

1914 Austrian archduke Francis Ferdinand is fatally wounded in Sarajevo which leads to the beginning of World War I. The Austro-Hungarian Empire fights on the side of Germany and the Turkish empire.

1918 After the defeat of the Germans by the Western Powers, the Austro-Hungarian Empire comes to an end and chaos spreads throughout Hungary. The peace treaty signed after the war divides much of Hungary's territory.

1938-1944 Hungary becomes an ally of Nazi Germany in World War II. Later regretting this decision, she tries to break with Germany and is invaded by the German army. Hungary is liberated by the Soviet army at the end of the war.

1949 The Hungarian People's Republic is established. The communist government bans free speech and travel. The Catholic Church is persecuted and Cardinal Mindszenty is sentenced to prison.

1954 Hungary joins the Warsaw Pact, a military alliance with other eastern European communist countries.

1956 The Revolution of 1956 takes place in October. Harsh repression by the Soviet Union follows when they invade the country. János Kádár becomes the new communist leader and remains in power for many years instituting a milder form of communism.

1980s The economy stagnates and inflation rises considerably.

1980 The first Hungarian astronaut is launched in the Soviet spacecraft *Soyuz 36*.

1983 A new electoral law is instituted requiring a minimum of two candidates for all elections.

1988 Kádár, the head of the Communist party, is ousted and laws are passed allowing for a multiparty democracy.

1989 On October 23 acting-President Szuros declares the new Republic of Hungary and officially ends the communist reign in Hungary.

1990 Hungarians vote in free elections for the first time in 40 years. József Antal of the center-right party is elected prime minister beating the communists who come in fourth.

1998 The Hungarian people elect a 35-year-old, center-right, Viktor Orbán, as prime minister preparing the country for the 21st century.

1

In Central Europe

If you look at a map of Europe you will see in the center of the continent a land-locked country that has an oval shape. This country is called Hungary. It is one of the smallest countries in Central Europe and has an area of 36,000 square miles (93,036 square kilometers). Half its population of 10,600,000 live in towns and the majority of them belong to the Roman Catholic faith. All Hungarian citizens, whatever their creed or race, have equal rights and opportunities.

The country is divided by the Danube River into two halves: east and west. An immense plain, flat as a pancake, sweeps away to the horizon to the east of the river. This is called the Great Hungarian Plain, or *Puszta*, which means "waste land." There are few bushes or trees in the *Puszta* except on the banks of the rivers and around the villages. The soil is salt and sandy in places and only fit for grazing. In other places it is rich farmland which has been turned into one of Europe's granaries.

Thousands of years ago much of Hungary was under water, covered by a shallow sea linked to the Mediterranean. Then

A typical small church in the north of Hungary. Most Hungarians are Roman Catholics.

earthquakes, wind, and rain gradually raised the level of the land. The sea became a shallow lake dotted with islands. Over the ages, deposits of mud, gravel, and sand (brought down by torrents from the surrounding highlands) filled in the lake, forming the Great Hungarian Plain.

Northward, the Great Plain merges into the foothills of the Carpathian Mountains which stretch in an arc beyond the Hungarian border. The hills were once covered in thick woods. Now they are bare and terraced in places. These terraces are planted with vineyards which produce the famous Tokaj wines.

Lake Balaton, the largest lake in Central Europe, is famous for its health-giving warm waters. It lies in the western half of the country in the area called Transdanubia. The lake is long and narrow, almost 48 miles (77 kilometers) long and only about nine miles (14 kilometers) wide. Along the northern shore it is about 10 feet (three or four meters) deep. North of Lake Balaton lies a high plateau covered in trees. It is known as the Bakony Forest. South of the lake is a lower plateau and two rocky hills of granite surrounded by fertile valleys and orchards. This area is also rich in minerals.

The people who live in Hungary call themselves Magyars, just as the people who live in the United Kingdom call themselves British. Other people–Germans, Serbs, Slovaks, and Romanians–also live in Hungary. They are called "ethnic minorities" because they account for only a small percentage of the population. There are Hungarian gypsies, too, who have lived in Hungary for many centuries and have darker faces than the gypsies seen in Western Europe. Strangely, the Magyars are not related to any of their neighbors and so form a race apart. But they are related to the Finns of Finland from whom they were separated over a thousand years ago, by migration and invasions from Central Asia and the Volga River region in Russia.

The Hungarians pay for their goods with a monetary unit called a *forint.* The *forint* is a decimal currency and is divided into one hundred *filler.* A few *forints* will buy a box of matches or an illustrated postcard. The Hungarians also use the metric system for their measurements and weights.

About 70 years ago Hungary was mainly an agricultural coun-

A view of fertile agricultural land in northwestern Hungary.

try and most of the people were poor peasants and herdsmen. They lived in hamlets and villages of whitewashed, mud-brick cottages. The men wore white shirts and tight black trousers and black hats with brims turned up all round. In cold weather they also wore shaggy sheepskin cloaks. The women went about their work in long, black, cotton dresses and white or black head scarves. They grew wheat and potatoes, and reared pigs, horses and cattle for food. Tobacco was grown locally and the orchards provided fruit.

During the last year of the Second World War (1939-1945), Hungary, like other countries in Europe, became a battlefield. There was bitter fighting between the Hungarian and German armies on one side and the Soviet and Romanian armies on the other. Towns and villages were bombed and shelled until they were turned into heaps of rubble. Finally, the Soviet army overwhelmed them and drove the Germans out of the country while the Hungarians signed a cease-fire agreement. When the war ended food was scarce in many parts of the country, and life was hard.

Much of this has changed because new factories and power stations have been built all over Hungary. Young peasants went to the towns and became factory workers. Those left in the country work on farms and are paid wages.

The towns, too, have changed. Every year new housing estates and apartment blocks were built. New sites on the edges of towns were cleared to make way for the building of mills and factories, parking lots and supermarkets. Hungarian scientists helped develop new computer systems and precision instruments. They also did research in connection with nuclear power stations and space exploration.

New housing being built in the town of Mako.

The Hungarian countryside has also changed. New highways have been built for the increasing traffic. More and more long-distance trucks now pass through Hungary. High-tension cables connect the Hungarian power-grid to those of neighboring Slovakia and Russia. Oil and gas are being piped from abroad to provide energy for industry.

Hungary is a colorful and unusual country which lies in the heart of Europe. The Hungarian people have lived for many years under foreign occupation and oppression, through terrible wars and revolutions. They have suffered famine and destruction. They are a proud people who have survived and preserved their traditions and are very much a part of the modern world. It is easy to travel to Hungary by many highways and railways. In addition, there are direct flight connections between Budapest, the capital, and many cities of the world. Let's visit this country and look more closely at the way of life of the Hungarian people.

2

Land and Climate

Hungary lies in the Middle Danube Basin in Central Europe. The country is surrounded by mountains which stretch in lofty ranges beyond its borders. These are the Alpine and Carpathian Mountains in the west and north, the Dinaric Mountains in the southwest and the Eastern Alps in the east.

As you will see from the map Hungary's neighbor in the northeast is the Ukraine. In the north Hungary borders on Slovakia and shares with it a stretch of the Danube River. Hungary's neighbor in the west is Austria. In the south it is separated from the countries of Slovenia, Croatia, and Serbia by the Dráva River which flows into the Danube from the west. Hungary's neighbor in the southeast is Romania. The border here runs mostly across the plains.

The eastern half of Hungary consists of the Northern Mountains and the flatlands to the south, known as the Alföld, Puszta, or Great Hungarian Plain. The Northern Mountains have two ranges–called the Mátra and Bükk Mountains–composed of hard volcanic rock and granite. The highest peak in all Hungary is in the Mátra Mountains. It is Mount Kékes which rises to 3,350 feet (1,015 meters). The highest point in the Bükk Mountains is Bálvány Peak which rises to 3,136 feet (956 meters). A national park has been established in the Bükk Mountains where ancient trees and waterfalls can be admired. Some of the rock here is limestone which has been worn into deep hollows and gullies by rain and wind.

The Aggtelek Mountains, which Hungary shares with Slovakia, are composed of limestone, sandstone, and other rock. They have pointed ridges and are bare and rocky. This is karst country. (Called after the Karst region on the Adriatic coast in former Yugoslavia.) There are many potholes, long caves, and grottoes

A view of the Mátra Mountains in northern Hungary.

Part of the Dripstone Cave of Aggtelek.

hidden underground. These caves are often connected to each other by narrow passages. The most famous is the Dripstone Cave of Aggtelek which is 13 miles (22 kilometers) long. In this ice-cold cave, visitors can see stalactites hanging from the roof and stalagmites rising from the floor in fantastic shapes.

In the upper reaches of the mountains, vegetation is sparse; only moss-like plants, gentians, and junipers grow well. Lower down, there are forests of oak and hornbeam, followed lower still by linden and ash. The forests were once full of wildlife. Now wolves and wild boars are almost extinct but there are still some deer, squirrels, and wildcats climbing the trees.

The Great Hungarian Plain covers an area of about 20,000

square miles (52,000 square kilometers). The plain itself is formed of layer upon layer of sedimentary rocks which go down to a depth of about 1,000 feet (300-400 meters). These layers of dust and grit were blown by the winds after the Ice Age and then trapped by the roots of grass. This type of soil is called loess, after a town in eastern France where it was first found. Loess stretches in a long belt from the plains of northern Europe, through Hungary and into the Ukraine where it can be recognized in the fertile black earth.

The Tisza River flows through the middle of the plain. It is a sluggish river with many bends, or meanders. Groves of white willow, hawthorn, and tartar maple line the river. At one time, the river flooded large areas of the plain each spring but today there are long stretches of embankments which protect the plain and the towns. In addition, dams have been built across the Tisza to regulate the flow and to help with irrigation. Farmers here grow rice, cotton, and maize and there are also many market gardens.

Away from the river, large stretches of the plain consist of poor grassland. Rye-grass and wild clover grow here, as well as weeds, camomile, and thistles. There are also sand-dunes which shift from place to place when the strong polar winds from the east blow across the treeless landscape.

The western half of Hungary consists of the Little Plains in the north and Transdanubia in the south. The Little Plains border on Lake Fertö in the northwest. Half of Lake Fertö lies in Hungary and the other half is in Austria. The lake is shallow and its shores are marshy and covered with reeds. The Little Plains consist of

Rushes stacked for drying.

thick layers of loess (fine yellowish-grey loam). Arable farming is carried out there on a large scale. Wheat, sugar beet, maize, and potatoes are the main crops.

Lake Balaton or the "Hungarian Sea" lies south of the Little Plains. It covers an area of about 232 square miles (600 square kilometers) and the temperature of the water reaches 77 degrees Fahrenheit (25 degrees Celsius) in summer. The southern shore is so shallow that you can wade into the lake only knee deep in water for about half a mile (nearly a kilometer). Near Lake Balaton, at Kápolnapuszta, is a unique buffalo reserve.

South of Lake Balaton is Transdanubia, once the Roman province of Pannonia. The Mecsek Mountains, with their broad valleys and forests of beech and oak, form the center of the region. Vineyards grow on the southern slopes of the mountains. There are also rich deposits of bauxite here, while coal is mined in the south. The Zala oilfields are found in the southwest. Some years ago, uranium was discovered in the Mecsek Mountains. The mining of this uranium was developed under strict control by the Soviet Union, because nuclear bombs could be made from it. Enriched uranium is also used at atomic power stations.

Hungary is situated between the latitudes of 45 and 48 degrees North which means that its climate belongs to the warm temperate belt. The western half of the country is rainier than the eastern half with an annual average rainfall of 24 to 32 inches (600 to 800 millimeters). July and August have plenty of sunshine. Hungary boasts an average of 2,000 hours of sunshine each year. The mean temperature in July varies between 64 and 73 degrees Fahrenheit (18 and 23 degrees Celsius). The earth is parched during these months and the countryside looks lifeless. By contrast, the coldest month of the year is January when the mean temperature varies between 32 and 25 degrees Fahrenheit (0 and minus 4 degrees Celsius). Most of the rivers and lakes are frozen then, including Lake Balaton, and thick snow covers the country, forming drifts in many places.

3

On the Banks of the Danube

Hungary's most important river and lifeline is the Danube. The Hungarians call it the Duna River, just as the Germans and Austrians call it Donau. The Danube is the second longest river

Pest, seen across the Danube. The river is an important trade route.

in Europe. It is 1,770 miles (2,848 kilometers) long. It flows through Hungary for nearly 258 miles (415 kilometers). Its entire drainage area (the area covered by the river and its tributaries) is nearly four times as great as England, Scotland, and Wales put together, or about one-tenth the area of the United States.

The Danube rises in the Black Forest in Germany, flowing in an easterly direction like a wild mountain river. But by the time the Danube reaches Hungary it has slowed down and become a wide river. At the border, the river divides into two branches, one called the Danube and the other the Mosoni Danube. Before the two branches unite further downstream, the Mosoni is joined by the Rába tributary. The main stream of the Danube forms the border between Hungary and Slovakia over a stretch of 62 miles (about 100 kilometers). Then it turns suddenly in a southerly direction. This point is known as the Danube Bend. South of Budapest, the Danube again divides into two branches which enclose the industrial island of Csepel. Beyond Csepel, the river continues in a southerly direction through flat, rich farmland. Here the water becomes sluggish, and islands and sandbanks form in it. White willow and black poplar groves line the banks. Great white herons and redshanks live among the reeds on islands and in marshes. On warm summer evenings, the air is filled with thick swarms of mosquitoes and the croaking of countless frogs can be heard all around.

Horses and large herds of long-horned cattle graze the meadows and use the river as watering-places. The extensive farms rear pigs. Some of these farms also have small herds of buffaloes which originally came from India. The buffaloes are black, with curved flat horns. They are gentle animals which like to seek shallow parts of the river or muddy oxbow lakes and wallow in the mud for hours, as pigs do. The farmers use them (instead of oxen or horses) for pulling carts. They also make cheese from the rich buffalo milk.

Because the Danube in Hungary has two regular floods each year–in early spring and early summer–a large area of the countryside used to be flooded. Today, there are nearly 1,900 miles (nearly 3,000 kilometers) of embankments along the Danube to protect farmland and settlements from floods.

The first Hungarian atomic power station was built at Paks, on the banks of the Danube south of Budapest, with Soviet help. The station began operating in 1982.

About 110 miles (178 kilometers) from Budapest, the Danube flows past the historic city of Mohacs in southern Hungary where two major battles were fought against the Turks in the Middle Ages.

By the time the Danube reaches the Croatian-Serbian border and flows on towards Belgrade, a strange thing has happened: the river has grown smaller in size. The volume of water carried by the Danube between Budapest and the border has diminished because the dry, flat banks absorb more water than the few tributaries bring. The only important tributary along this stretch of the river is the Tisza, which joins the Danube south of the border with Croatia and Serbia.

The Danube and its tributaries mean a great deal to Hungary.

An embankment along the Danube, built as a protection against flooding.

They provide freshwater fish, and water for livestock and irrigation. Because of the irrigation, Hungarian farms produce rich crops of tomatoes, green peppers, cucumbers, and other vegetables.

In the Middle Ages, when there were few roads and no railways, the Danube was an important line of communication between Western Europe and the countries on the shores of the Black Sea. The great flow of trade from the Turkish Empire and the Middle East came into Europe that way. For most of the length of the Danube, navigation is easy, except in winter when the river carries dangerous ice-floes. There is a regular fast passenger-boat service between Budapest and Vienna from April to October. The boats used are hydrofoils which skim over the surface of the water.

The river is also used for the transport of heavy and bulky goods. But river transport is slow, and today, when roads and railways have been built everywhere and important cargoes are even carried by air, the river steamers and barges are at a disadvantage. In spite of this, the Danube is still an important trade route, especially for such goods as cement, gravel, oil, and timber.

A special radio service gives daily news of the river's levels to help the pilots of tugs and steamers with navigation. Because much of the traffic on the Danube is international, the news on the radio is given in several languages.

If we look at the traffic on the Danube from one of Budapest's quays, we see tugs at work towing strings of steamers laden with girders, cables, and tractors. The river steamers are nearly 100 feet (30 meters) long, smaller than a sea-going ship. The crews of the various boats, who know each other well, often greet passengers with the song they all know:

The waters of the Danube are broad and strong
The waters of the Danube run laughing along.

Anyone who has traveled down the river has no difficulty in understanding how important this great waterway has been for many centuries, and still is, to all the countries through which it flows. The history of the Danube countries has been a stormy one, with many wars. So, in 1856, after the Crimean War, the statesmen of the European powers signed a treaty which stated that the Danube should be a free highway to the ships of all countries. They set up an International Commission to regulate navigation, settle disputes, and keep the river clear by dredging channels of sand and silt. In 1948, after the Second World War, this Commission was replaced by a new authority called the Danube River Commission. The Commission has its headquarters in Budapest and its members today are all the countries (with the exception of Germany) through which the Danube passes.

4

Early History and Kings

About 3,000 years ago, the Hungarians–or Magyars–lived on the banks of the Volga River in the plains of present-day Russia. They were known as the Ugrians, and they were short in stature with black beards and deep-set eyes. The women were skilled in pottery, weaving, and spinning. The men reared livestock (mainly cattle and horses) and tilled the land with wooden plows. They were good fighters with spear and arrow; they had to be, because hordes of nomadic people swept across their lands from time to time.

In the fifth century A.D., a host of Mongolian horsemen called Huns, led by their cruel chieftain Attila, roamed the plains of Europe destroying settlements and killing people. A branch of the Huns known as the Khazars conquered the Hungarian tribes and took over their lands. The Khazars (who were also rich merchants) eventually became overlords of the Hungarians. Though the word "Hungarian" is derived from the name Hun, the Hungarian people are not direct descendants of the Huns. Nevertheless, Attila is a common name among Hungarian men.

In the meantime, present-day Hungary was mostly waste land. The Romans had crossed the Adriatic in the third century B.C. and conquered the Illyrians who lived in the land that is now part of Yugoslavia. The Romans then conquered the lands to the west of the Danube, today known as Transdanubia, and turned them into the province of Pannonia. Beyond the Danube, the Romans clashed with the Dacians in the first century B.C. They later conquered the Dacians and turned the country into the province of Dacia.

The remains of the Roman town of Gorsium.

In the following centuries, the Roman legions withdrew from both Dacia and Pannonia because of the attacks of the barbarians. The ruins of Aquincum on the outskirts of Budapest, as well as the ruins at Szmobathely and Pécs, are evidence that the Romans once lived there. Historians, however, have not been able to agree on how many people were left behind by the retreating Romans to work the land and make a living. Some historians even say that the land was left deserted and free to be occupied by newcomers from the north and east.

In the early Middle Ages, the Hungarian tribes who still lived far away to the east of the Carpathians found life more and more difficult and insecure. So the age of Great Migrations began. The tribesmen left the steppes in the second half of the ninth century A.D. under their chieftain Árpád. They made their way across rivers and woods, hills and valleys, and finally crossed the Carpathian Mountains. They pitched their tents made of pelts, and (by the end of the ninth century) settled on what was to become known as the Great Hungarian Plain.

The Hungarians were a warlike people and their archers were much feared. Their horsemen raided and plundered neighboring people as far afield as Bulgaria, Italy, Germany, and Poland. The German knights resolved to put an end to Hungarian raids. In 933 A.D., an army of German knights, led by King Henry V, fought and routed the Hungarian marauders on the battlefield of Merseburg. However, the Hungarians continued to be troublesome and so the German king Otto I led his knights at the battle of Lechfeld near Augsburg in 955 A.D. and finally defeated the Hungarian hordes.

A statue of King Stephen I, who became Saint Stephen after his death in 1083.

In 997, King Stephen I (King István), who was descended from Árpád, came to the throne of Hungary. He received the crown from Pope Sylvester II and reigned until the year of his death. King Stephen I united all the tribes of his country into one nation and also conquered Transylvania. He converted his people to Christianity and the Roman Catholic faith. For this, he was canonized (made a saint) by the Pope, after his death in 1083. King Stephen I is regarded as the greatest king of the Hungarians because he founded the Hungarian monarchy which lasted until the battle of Mohacs in 1526 when the Turks conquered the country. After that date, Hungary was not to be a free state again for centuries.

Another descendant of Árpád was King Andrew II who reigned from 1205 to 1235. He will always be remembered, because he went on a crusade to the Holy Land to please the pope in Rome. In 1231, King Andrew II signed a Golden Bull (charter) which gave tax exemptions and liberties to the descendants of Magyar noblemen.

King Andrew II was followed by King Béla IV who reigned until 1270. His reign is notable for the scourge of the Tartar invaders. The Tartars were also known as Mongols. They had slit eyes and were swift and hardy horsemen. They were led by Batu Khan and they invaded Hungary from the northeast in the spring of 1241. They killed the people and plundered the land wherever they went. They withdrew in 1242, as suddenly as they had arrived, and fled back to Asia. King Béla then built a stone fortress overlooking the Danube, known today as Buda Castle.

The last king of the House of Árpád died in 1301. Then the reign of the Angevin kings started. Charles of Anjou was elected king and he was followed by his son Louis. Another king, Sigismund of Luxembourg, reigned from 1387 to 1437 and was also crowned king of Bohemia. It was during his reign that the famous Protestant preacher called Jan Huss was brought to trial by the Roman Catholic Church and was found guilty of being a heretic. He was burned at the stake in 1415.

During and after the reign of King Sigismund, the Hungarians built many castles and forts, defended by thick walls with loop-holes and tall towers. Some of the castles stood high on rocky hills, while others were built inside river bends so that they were

Buda Castle, built by King Béla IV.

protected by water on three sides. The castles were built mainly because the people feared a Turkish invasion. The Turks had destroyed the Serbian army on the Field of Blackbirds in 1389 and were continuing to march northward. At first, there were skirmishes between parties of marauding Turks and Hungarian border guards. Then the Turks tried to invade the principality of Transylvania and were defeated by János Hunyadi, who was a powerful baron and skillful general.

Transylvania means "land beyond the forests." It is a plateau surrounded by mountains. The Hungarian kings allowed Szeklers (a people claiming descent from Attila's Huns) to settle in

Transylvania in the 12th century and they were followed by Saxons in the early 13th century. An old legend says that these Saxons were the descendants of the children who followed the Pied Piper of Hamelin after he rid the city of rats.

Transylvania has rich farmlands as well as gold and silver mines. It was these which attracted the interest of the Turks. So Sultan Murad II sent an army of a 100 soldiers to take Transylvania. This time, Hunyadi was defeated at the battle of Kossovo in Serbia, in 1448. (He died of the plague in 1456.)

Faced with the Turkish threat, the Hungarian barons elected Hunyadi's son, Matthias, as king of Hungary and Bohemia in 1458. Matthias Hunyadi (also known as Corvinus I) was a good general and he defeated the Turks in battle several times. He was an enlightened ruler who encouraged artists and builders from Italy's golden age of the Renaissance to come to Hungary where they built palaces and castles in the baroque style. Matthias Corvinus also built in Transylvania the beautiful Hunedoara Castle with its slender turrets.

At the turn of the 16th century, the Turks began preparations for the final conquest of Hungary. A major battle was fought, at Mohacs on the banks of the Danube, in 1526. The Turkish artillery and the assault of the infantry (or janissaries) won the day. The remnants of the Hungarian army, led by King Louis, fled. King Louis himself drowned while attempting to cross the Danube. The Turks occupied Buda and Pest in 1541. By 1547 they had conquered all southern and central Hungary and settled in the towns. They remained as rulers for the next century and a

Matthias Hunyadi (Corvinus I), who was elected king of Hungary and Bohemia in 1458.

half. Transylvania recognized the Sultan as overlord and agreed to pay tribute, or what we would call "protection money."

5

The Austro-Hungarian Empire

After the battle of Mohacs, Ferdinand I of Hapsburg became king of Hungary. He reigned until 1569 and, during his reign, Turkish and Austrian armies often clashed on Hungarian soil causing death and destruction among the local inhabitants. They plundered the people and levied taxes.

In 1569 the Hapsburg Emperor Maximilian II signed a treaty with the Turkish Sultan Selim II at Adrianople. Under this treaty Hungary was divided into three parts. The kingdom of Hungary remained in the north. The principality of Transylvania was established in the east, and central and southern Hungary remained under Turkish occupation. The Turks built many baths and mosques in Hungarian towns at this time.

However, Turkey wanted still more territory and, in 1683, a powerful Turkish army of archers and horsemen pushed northward and began the siege of Vienna. They were led by the Grand Vizier Kara Mustapha. The siege lasted a long time and the situation became desperate for the defenders. Emperor Leopold I asked the Polish king John Sobieski III for help. King Sobieski arrived with his cavalry of hussars and they crushed the Turkish army, at the walls of Vienna. The Turks fled and their commander lost his magnificent embroidered tent, his weapons studded with jewels, and his precious carpets. Three years later Buda was freed from the Turks. The Austrian army pursued the Turks southward; and finally, in 1699, a treaty was signed at Karlowitz in Serbia, and the Turkish sultan gave up his claim to Hungary.

A minaret in Pécs—a reminder of the Turkish occupation in the 16th century.

A statue of Ferenc Rakoczi, outside the Parliament Building in Budapest.

The Hapsburg emperors became the rulers of Hungary which became a province of the Austrian Empire. This caused much discontent, since heavy taxes were levied on salt, and purchase tax was introduced.

The Hungarian Prince Ferenc Rakoczi led a rebellion against the Austrians from 1703 to 1711. But, in the end, the Hungarian nobility accepted Hapsburg rule and the Treaty of Szatmár was signed in 1711. In 1723 the Hungarian *diet* (parliament) accepted a treaty which bound Hungary to Austria.

During the 18th century the borders of the Hapsburg Empire stretched from the Carpathians to the Adriatic Sea. It was one of the largest empires in Europe and it included the whole of Hungary and Transylvania.

The Empress Maria Theresa was crowned in 1740. She was an autocratic ruler and, during the 40 years of her reign, she levied many taxes on the Hungarian people, causing Hungarian patriots to continue the struggle against Austrian rule. This struggle continued into the 19th century. A famous Hungarian leader of the time, Lajos Kossuth, was born in Monok in 1802. He came from an impoverished noble family but had qualified as a lawyer. He waged a campaign during the years 1832 to 1836 demanding reforms. Kossuth was put in prison by the Austrians but later released. Then, in 1848, at a time when almost the whole of Europe was in revolt, he led a revolution in order to gain civil liberties for Hungarians. Kossuth was supported by the poet Sandor Petöfi who was considerably younger. Petöfi was born at Kiskörös in 1823. He died fighting the Austrians and has become

A portrait of Lajos Kossuth, the Hungarian revolutionary leader and patriot.

The military dispersing rioters in Budapest in 1848.

Hungary's national poet. The Russians assisted the Austrians in defeating the rebel Hungarians and so the war of independence of 1848 was lost. Kossuth ended his days in exile in Italy.

In spite of his success in crushing the rebellion, the Austrian Emperor Francis Joseph was forced to recognize Hungary as a self-governing kingdom. Thus the Dual Monarchy was born and, in 1867, the Emperor of Austria was also crowned king of Hungary in Budapest with the crown of St. Stephen. The new country became known as the Austro-Hungarian Empire.

At the turn of the 19th century, the Austro-Hungarian Empire ruled much of present-day Czech Republic, Slovakia, Yugoslavia, and Transylvania (which now belongs to Romania). The peoples living in these countries wanted independence but the Austro-Hungarian rulers were determined to prevent this as they did not want to lose their territories, titles, and privileges.

When the Austrian Archduke Francis Ferdinand visited Sarajevo, the "city of palaces," in 1914, he was shot and fatally wounded by a Bosnian student who wanted freedom for his country. The archduke's death led the Austro-Hungarian Emperor to declare war on Serbia. And this was the beginning of the First World War.

Great Britain, France, Italy, and Russia joined the war on the side of Serbia. The Kaiser's Germany and the Turkish Empire supported the Austro-Hungarian Empire. At first, the Austro-Hungarian troops won many victories. They defeated the Russians and beat the Italian army at Caporetto on the Isonzo River. Then Austro-Hungarian troops, together with German units, crossed the Carpathian Mountains and conquered much of Romania. But after that the Western Powers started to win battles. The Turkish Empire collapsed in 1918. The war ended in defeat for Germany. And the Austro-Hungarian Empire came to an end. At last, the nations which formed the empire, including Hungary, were given the right to decide their own fate and so gain independence.

6

Recent History

During the closing stages of the First World War, Hungary was ruled by King Charles IV who reigned from 1916 to 1918. The end of the war brought to power in 1918 a Hungarian National Council headed by Count Mihály Károlyi. But chaos spread

Trucks full of rebel troops are cheered in the streets during the first stages of the Hungarian democratic revolution in 1918.

throughout the country. Food was in short supply and many people were unemployed. In the winter of 1918, revolutionary workers led by the communist Béla Kun (who came from the Soviet Union) proclaimed the Hungarian Soviet Republic. As a result of this situation, Czechoslovak and Romanian troops invaded Hungary and occupied Budapest. The revolution was crushed in 1919. A new kingdom of Hungary was proclaimed without a king. Instead Admiral Nicolas Horthy was appointed regent in 1920. However, this was not the end of the Hungarian people's troubles.

The peace treaty between the victorious powers and Hungary was signed at Trianon, in France. Hungary lost much territory and many of her citizens found themselves in newly-formed countries such as Czechoslovakia, Yugoslavia, and Great Romania. Some historians believe that a great injustice was done to Hungary and that it was punished harshly because it fought on the losing side in the First World War. In Hungarian towns visitors can still see monuments with the inscription "Justice to Hungary."

Hungary became the ally of Hitler's Germany before the Second World War, seizing the chance to take back some territory when Czechoslovakia was partitioned in 1938 and 1939. In 1940, an agreement was signed in Vienna called the Vienna Arbitration Award. This gave back to Hungary a large part of Transylvania which had belonged to Romania. Part of the Romanian population was rounded up, put in cattle-trucks and sent to Romania. When Nazi Germany invaded the Soviet Union in 1941, Hungarian troops joined the Nazi army. In December 1941, Hungary declared war on Britain and the United States.

However, things went badly for Hungary in the war, and Admiral Horthy tried to break the alliance with Germany. Now Hungary was taken over by the Nazis and Horthy was deported to Germany. After bitter fighting, Hungary was liberated by Soviet and Romanian troops in January 1945 and signed an armistice which meant giving up all territory acquired with German help. So Hungary was now restored to its pre-1938 frontiers.

The country became a republic in February 1946. Land was taken from the big landowners and distributed to peasant cooperatives, and industry was nationalized. In the 1947 elections, the communists became the largest single party. The new Hungarian National Assembly ratified the peace treaty on July 2, 1947. The treaty allowed the stationing of four Soviet divisions in Hungary. The Hungarian People's Republic was established in 1949.

Hungary joined the Council for Mutual Economic Aid (Common Market) of the communist countries in 1949. Then, in 1954, it joined the Warsaw Pact–which is a military alliance of communist countries, similar to the western defense system called the North Atlantic Treaty Organization (NATO).

When the communist government of Hungary came to power it decided to retain the country's traditional flag which has three equal horizontal stripes–red, white and green–but to discard the royal coat of arms which used to appear in the middle of the white stripe. A new coat of arms was adopted consisting of a red five-

The Hungarian flag, with its three equal horizontal stripes of red, white and green.

pointed star surrounded by sheaves of wheat, and, below the star, a golden hammer crossing an ear of wheat.

The communist government banned free speech and travel abroad and the secret police was much feared by the people. The Catholic Church was persecuted. Its head, Cardinal József Mindszenty, was tried in a people's court and sentenced to life imprisonment.

General poverty and low wages in postwar Hungary caused discontent among the workers. There were demonstrations against the government in the autumn of 1956. The police fired on the crowd of demonstrators. The angry crowd started an armed uprising and the army joined the insurgents. Members of the secret police were hunted down, dragged from their homes, and executed in the streets. Communist books and papers were put on bonfires. A new government came to power headed by Imre Nagy who appealed to the United Nations for protection against Soviet interference. The Soviet Union reacted swiftly by sending troops and tanks into Hungary. Soviet planes bombed Budapest. Street-fighting took place in the capital and other cities, and the Soviet army won the day. About 150,000 Hungarians fled to the west by crossing the borders into Austria and Yugoslavia.

A new pro-Soviet government was set up, headed by János Kádár who also became leader of the newly-formed Hungarian Socialist Workers' Party. János Kádár was party leader and the most powerful politician in Hungary until 1988.

When economic reforms were introduced in Czechoslovakia in the 1960s, the leaders of the Soviet Union, Hungary, East Germany, Poland, and Bulgaria met in Warsaw and from there sent an ultimatum to Czechoslovakia to put an end to these reforms. As no satisfactory answer was given to the ultimatum, during the early hours of August 21, 1968, Czechoslovakia was invaded by troops and tanks from Poland, the Soviet Union, and Hungary. That was the only occasion, since the end of the Second World War, on which Hungarian troops have invaded a foreign country.

In May 1980 the first Hungarian astronaut, Squadron Leader Bertalan Farkas, was launched on board the spacecraft *Soyuz 36* on a joint space mission with the Soviet Union. He spent a week orbiting in space before he came back to earth in Central Siberia.

In the same year as this historic event took place, Hungary achieved a surplus in its balance of payments with the west–the first since 1973.

The 1980s saw the economy go into stagnation and inflation rose rapidly. The people of Hungary became disallusioned with the communist government and in 1988 began a peaceful transition to democracy with the ouster of Kádár. In 1990 the Hungarians went to the polls in the first free election held in 40 years.

7

Industry and Transportation

At the beginning of this century Hungary was mainly an agricultural country. Most of the people were peasants, some of whom were poor and worked for the big landowners. Since then, Hungary has become an industrialized country, with heavy and light industries. Today Hungarian goods are exported all over the world.

The closing stages of the Second World War brought ruin to industry and transportation in Hungary. The first task of the government after the war was to rebuild factories, bridges, and railway lines. Economic development was planned by experts in Budapest. They introduced the Five-Year Plans, each one of which marked a stage in the country's development. The fourth Five-Year Plan, for instance, which lasted from 1971 to 1975, greatly increased industrial production. The fifth Five-Year Plan (from 1976 to 1980) laid emphasis on modernizing industry. The sixth Five-Year Plan (from 1980 to 1985) introduced economic reforms and a drive for making industry efficient and profitable.

In 1982, Hungary joined the International Monetary Fund (IMF) and was granted a big loan for developing industry. In July 1983, Hungary and the World Bank reached an agreement on the provision of loans for developing the energy industry and the production and storage of wheat. Unknown in other east European countries, several western firms were allowed to become partners of Hungarian firms in order to develop the country's resources. New branches of industry were created in Hungary. Agricultural equipment, buses, electrical goods, plastic materials, and fertilizers are now being manufactured. Most of Hungary's industry is concentrated on the banks of the Danube, and in the north and west of the country. Farming still predominates in the east.

One of Hungary's main industrial areas is Budapest and its surroundings, including the island of Csepel to the south, located between the two branches of the Danube. Budapest has heavy and light-engineering works, chemical and textile factories, and industries specializing in processing leather and fur, and the production of electrical goods. The island of Csepel has iron and electronic goods. and steel mills, engineering works, paper factories, and an oil refinery. The island is also a free port for the international Danube traffic and has busy docks along the riverfront.

Hungary's third largest city, Miskolc, is in the northeast of the country. Miskolc is a sprawling industrial center with good road and rail connections. It had many factories making textiles, glass, cement, furniture, and food products, as well as engineering works but many of these now lay idle. Its merchant houses trade in local wines and tobacco.

Factories in the town of Ozd. After the Second World War, great emphasis was placed on expanding industry.

Debrecen, which is Hungary's second largest city, lies on the edge of the Great Hungarian Plain and is an important center for trade and handicrafts. There are factories making furniture and paper, and food-processing plants where the famous Debrecen sausages are made.

Pécs, another large city, lies near a coal-mining area in southern Hungary. It has breweries, distilleries, and a famous ceramic factory. The land around Pécs produces tobacco, fruit, and vegetables, and there are extensive vineyards dating back to Roman times.

Szolnok is one of Hungary's smaller towns but, because of its central situation, it has good lines of communication with all parts of the country. Szolnok has factories producing cellulose, alcohol, sugar, and flour.

Györ, which is situated in northwestern Hungary, lies on one of the main trade routes with the west. It has factories making clothing and footwear, railway rolling stock, and diesel locomotives. Its warehouses handle much of Hungary's exports to Austria, including meat, fruit, vegetables, and wine.

The first railway in Hungary was opened in 1846 and it linked Pest to Vác. Today there are 4,729 miles (7,611 kilometers) of track, of which 1,379 miles (2,207 kilometers) are electrified. Hungary contains some of Europe's most important railway junctions–Budapest is linked by rail to Prague, Vienna, Belgrade, and Bucharest.

There are 42,730 miles (68,370 kilometers) of main and secondary roads but only 260 miles (420 kilometers) of highways. Most Hungarian roads are straight, with few bends. They are shaded by poplars and cherry trees from the fierce summer sun. Hungary lies on the main motor haulage routes from London to Damascus in the Middle East and from Hamburg in Germany to Bucharest. Hungarian traffic rules are based on the Geneva and Vienna international conventions. People drive on the right-hand side of the road and there is a speed limit for cars of 37 miles (60 kilometers) an hour in built-up areas. The speed limit on roads is 50 miles (80 kilometers) and on highways 62 miles (100 kilometers).

Hungary also has 1,048 miles (1,688 kilometers) of navigable inland waterways, the Danube being the most important. There are regular boat-services from Budapest to the Danube Bend, run by the Hungarian Shipping Company (MAHART). There are also boat-services and ferries across Lake Balaton. In addition, Hungary has a small merchant fleet.

International communications include the Hungarian airline MALÉV, which operates from the country's only international airport at Ferihegy, about 10 miles (16 kilometers) from Budapest. It has an extensive network of international flights but no internal flights. Easy transportation to and from Hungary has stimulated the growth of the tourist industry. Restoration is being done to many of Hungary's magnificent castles for use as resorts. Most tourists come from neighboring countries, such as Slovakia and Austria; but considerable numbers also come from all over the world.

8

Forests and Fisheries

Forests do not grow well in the plains of Hungary so the country has to import soft timber for house building, telephone poles, and other industrial purposes. Only 19 percent of Hungarian territory is covered by forests, though in the last century the woodland area was much larger.

There are no natural coniferous forests except for a few fir and larch plantations. In the Northern Mountains there are extensive forests of beech and oak which can reach a height of about 80 feet (24 meters). The beech is used for making furniture, while the oak is much sought after for parquet floors. Oak is also used for making barrels for the beer and wine trade. Ash and hornbeam grow well here, too. Hornbeam is a small tree with a twisted trunk and smooth grey bark. In some places, the leaves of the trees form a thick ceiling so the sun can hardly shine through. Wildcats, boars with big tusks, and even wolves have their hunting grounds there. But there are also forest clearings where the timber has been felled. Here the ground is rough and steep, and

A forest near Miskolc in northern Hungary.

timber haulers remove the logs using teams of heavy Nonius horses. When the timber reaches the road, the logs are loaded onto trailers drawn by powerful diesel trucks.

About one-third of the timber is used for firewood since many houses, especially in the countryside, have wood-burning stoves. But some gnarled trunks and branches are left behind for the charcoal-burners. This wood is piled into a tight mound, covered with turf and set alight to burn slowly inside, without flames, until the wood is charred black and becomes charcoal. The freshly-made charcoal is then put in bags and sold to inns and hotels for cooking the famous Hungarian barbecue dishes and grills.

The straight timber that has been felled goes to sawmills to be cut up into deals, plywood, and other useful wood. Hungarian carpenters and cabinetmakers are skilled at carving table legs and chairs. They also make whole furniture suites covered in veneered walnut–the walnut has a beautiful grain and is cut into thin sheets which overlay a cheaper wood.

In the last few years a major reforestation program has been carried out to preserve the woodland areas and replace the timber which has been taken away. Young trees from nurseries, mostly oak and beech, are transplanted in the autumn and fenced in for protection from rabbits and deer.

Commercial fishing can only be carried out on a small scale in Hungary, because the country has no seashore. The most important freshwater fish is caught in the Danube and Tisza rivers and includes cod, pike, perch, and smelt. Catfish, so-called because of its long feelers resembling a cat's whiskers, is also caught. Carp is another popular freshwater fish. It is bred in lakes and fishponds where hogbeam, marsh fern, and white water lilies grow. Rainbow trout is caught in the mountain streams.

The river fishermen work from wooden platforms on the banks, or from small boats. They use nylon dip-nets spread out at the end of long poles which they lower into the water. Their catches today are not as good as they once were, because fish have died in large numbers due to pollution and the discharge into the water of poisonous chemicals from factories on the banks of the Danube. In their spare time some fishermen cut osiers–the shoots of willow trees which grow on the riverbanks–and make wicker chairs and tables, as well as baskets of all shapes and sizes.

9

Farms and Crops

Today only one-eighth of Hungary's population are employed in farming as opposed to half of the population only a few years ago. Farms have always produced enough food for the country's population plus a surplus for export. The farms also supply manufacturing industries with raw materials such as beet for the production of sugar, or hides for making shoes and other leather goods.

Before the Second World War, most farms in Hungary were large estates belonging to the nobility. The estates employed farmhands to do all the work. For themselves, the farmhands had only a small patch of land on which they grew a few vegetables and fruit trees. They also kept hens and geese. The Communists, coming to power after the war, changed the whole aspect of Hungarian farming. Under Communism, about 94 percent of farming land was owned by the state or cooperatives. Since 1990 farmland has been reprivatized and privately-owned farms are the norm now.

Agricultural output varies from year to year, according to the weather. The eastern half of the country is drier than the western half and here crops–and the food supply–can be affected by drought. For example, Hungary had a poor harvest in 1983. As a result of this, some basic food prices were increased sharply. Irrigation of crops has become of vital importance. More and more irrigation schemes have been created using the plentiful waters of the Danube and Tisza rivers. Some of the irrigation is furrow irrigation where the water runs into fields along channels. But spray irrigation, involving the use of pumps and pipes, is being used more and more for fields which are far from river banks.

Traveling in summer through farming country near Szeged the visitor sees fields of wheat, barley, and corn. The wheat ripens by early July so the combine harvesters are busy at work throughout the month. Corn (maize) ripens by mid-October. It is an important crop because it not only provides food but is also used for making starch, glucose and alcohol. In addition, the dried stems and leaves of the maize plant are used as fodder for cattle and sheep.

The sunflower is also an important crop popular with Hungarian farmers. The sunflower plants grow to a height of over six feet (two meters) and their seeds contain a high proportion of oil–sometimes as much as 35 percent. The oil has a clear pale yellow color and is used as a salad oil, for cooking and for making margarine.

Sugar beet is grown on the arable lands of northwestern Hungary. It is one of the country's most profitable crops. The tops of the sugar beet are used for cattle food. The pulp, which is returned to the farmer from the factory after the sugar has been taken out of the root, is also used for feeding cattle. A few years ago sugar-beet harvesting was done by hand and was a back-breaking job. Today, machines are used to top, lift, and deliver the beet into a tractor-drawn trailer moving alongside.

The hot summers and good supplies of irrigation water make the plains near the Danube and Tisza rivers suitable for growing rice. Tobacco plants are also grown in the sandy soils of southern Hungary.

Fields of cultivated poppies can be found in all parts of the country. Poppy seeds are left to ripen and, when collected, are used for sprinkling on bread loaves or rolls. Mixed with sugar, the seeds make a tasty filling in cakes. White mulberry trees also grow in many parts of the country; the berries are used for mak-

Sunflowers grown for the oil contained in their seeds. They are widely cultivated in Hungary.

ing jam. The leaves are fed to silkworms. When the silkworms are mature, they are placed in boxes lined with oak leaves and twigs, and among these they spin their cocoons. The golden cocoons are then collected before the moths have had time to develop and spoil the threads. Next, the cocoons are baked lightly in an oven. Finally, they are placed in boiling water. The threads thus become loose and are wound off easily. Several threads are twisted together to make silk yarn which can be woven into natural silk cloth.

There are many orchards of apricot trees in the valleys south of Lake Balaton. The apricot tree blossoms early in the year–at the beginning of March–so an early frost-free spring is essential. The apricots ripen by July. Some of the fruit is used by the canning industry for making jam and some is sliced and dried. Over-ripe fruit is put in vats and left to ferment. The fermenting pulp is sent to a distillery and turned into the famous Hungarian apricot brandy.

A large variety of apples, pears, and cherries are also grown. Most of the fruit is sold locally or sent to jam factories. Some farms specialize in growing quinces. The quince is an unusual fruit; shaped like an apple, it is hard and bitter. However, it makes good preserves and mixes well with stewed apples.

Hungarian market-gardeners grow vast numbers of small cucumbers (called gherkins) suitable for pickling. They are pickled in a mixture of water and vinegar, flavored with salt, sugar, and spices. Jars of pickled Hungarian gherkins are exported and can be bought in supermarkets all over Europe. The farmers of Hungary also grow herbs, like thyme, sage, and mint, because they are fond of spiced food. They also grow caraway seeds which are used to flavor bread and to mix with pickles. Medicinal herbs, too, are popular. The white flowers of the camomile, lime-tree blossom, and the stalks of cherries are dried and then made into herbal mixtures. Similarly, parsley roots are used to help relieve rheumatic pains.

The foothills of the Northern Mountains have good soil for vine-growing. The vineyards are planted on sunny slopes. The vines are carefully pruned and, from early June onwards, sprayed to protect the leaves and fruit from mildew and other diseases. The grape-picking season is always a festive occasion when many young people come to help carry the wicker baskets full of grapes. The Tokaj vineyards are highly productive and are famed for their fruity red wine. To the west, the vineyards at Eger are known for their Bikaver Bull's Blood wine which is exported to many countries of the world. There are also vineyards in southern Hungary; and others on the slopes of the Mecsek Mountains, from which light wines are made.

Hungary used to be a land of cattle. Because there were so many cattle the government in 1990 tried to combat the overproduction of livestock and a large reduction in the number of livestock has occurred. Since most of the land is not fenced, herds are kept in pens on the farms. When the cattle are driven out onto the plains in the spring, the herdsmen stay with them. They live in little wooden huts until they bring the beasts back to the farm in the autumn.

A view of the town of Tokaj, in one of the most productive vine-growing areas in Hungary.

There are also over one hundred thousand horses in Hungary. Some are still used on farms for pulling carts or trailers. Horses are also slaughtered for meat. Shagya Arabs are bred at Babolna, a stud farm established in 1789. These horses were originally imported into Hungary from Arabia, and crossed with nonpedigree mares. They stand 15 hands high and look very beautiful. Lipizzaners are also bred in Hungary. They are usually grey and have a sleek coat. They are intelligent and gentle and suitable for any kind of work. Another Hungarian breed is the Furioso, which usually stands sixteen hands high. The Furioso is black or dark brown and used in many equestrian sports including racing and jumping. The Nonius horses are a heavy Hungarian breed. Some stand up to 17 hands high and they make useful farm horses.

Hungarian farmers grow a lot of hay and red clover which they use as fodder for their cattle and horses. The red clover is a useful crop, because it can withstand hard winters. The farmers also grow fields of red clover, let it flower and dry out. They then mow the dry clover and put it through a combine harvester to extract all the seeds. Red clover seed is then exported to countries as far away as Canada.

The clover is one reason for beehives being a common sight in Hungarian village gardens. There are about 500,000 hives in

Cattle on the Great Hungarian Plain.

One of the famous Lipizzaner horses which are bred in Hungary.

Hungary. The bees also like visiting the yellow flowers of the acacia trees. Acacia honey is sold locally and also exported to many countries.

Another typical Hungarian crop is the red chilli, two varieties of which are grown in southern Hungary. The slender chillies are picked by hand and strung up on wooden frames to dry in the sun. Then the chillies are ground into a red powder, called paprika, which is very peppery and used to flavor goulash, stews, and other dishes.

10

Csikos and the Horse Round-up

In order to see what life is like in the plains of Hungary, let's visit a village in the eastern corner of the country on a summer day. The village is called Kisháza and is approached by a wide dusty road with some potholes in it. On the edge of the village there is a strange-looking well. It consists of a forked tree trunk, topped by a long pole with a chunk of concrete tied securely at one end. At the other end there is a long chain attached to a wooden bucket. The bottom of the well is some 10 feet (three meters) below the surface. On either side there are long wooden troughs for watering cattle and horses.

The village has a wide street of asphalt, with raised sidewalks of earth on either side. A few poplar trees with whitewashed trunks grow along the road. The village houses are low whitewashed buildings made of mud bricks. They mostly have thatched roofs but some are covered with wooden roof-tiles called shingles, which keep out the rain. Sheds and barns have roofs of corrugated iron. The village has a church, a school, a post office, two or three small shops, and an inn with a yard at the back surrounded by a privet hedge. In the middle of the yard there is a gnarled mulberry tree. There are also long tables and benches where customers drink beer or strong apricot brandy. Hot beef and onion stews are also served at these tables.

Inside the village houses there are wooden beds with huge pillows and duvets. There are also wooden tables and chairs and carved chests. Brightly-colored rugs and decorative plates hang on the walls. Above one bed there is a rosary of black beads, and a crucifix. At the back of each house there are a few apple and pear trees, a climbing vine full of black grapes, and a vegetable garden. The villagers grow their own vegetables including onions, garlic, tomatoes, and green peppers.

Hens and geese roam freely about the road. A goat with her kid is tethered on the grass verge in front of one house. In front of another there is a water-pump, as piped water is not available in this village. There are few telephones but there is electricity.

Outside the village gleam the white buildings of the collective farm which owns great herds of cattle and horses. Beyond the stables, barns, and sheds of the farm surrounded by a grove of acacia trees with sharp thorns, stretches the *puszta* or plain. There are herds of white long-horned cattle and horses, mostly chestnuts and greys, scattered on the treeless plain which shimmers in the hot sun.

As it is branding-day, several horsemen have ridden out across the plain to round up mares and their foals. They wear high black

A herd of horses on the Great Hungarian Plain, with a typical well in the foreground.

boots with spurs and carry long whips and lassos. The cowboys are called *csikos* (pronounced *cheekosh*) and they ride their horses as naturally as other people might walk. They keep in touch with one another with sharp blasts on a cow horn, and they soon find the herd. Then they shout and crack their whips and the herd gallops with wild eyes, sending up clouds of dust in their wake. The foals soon tire; several colts strike out on their own only to be driven back. Finally, they are all herded into the corral owned by the

A *Csikos*—the Hungarian equivalent of a cowboy—in his traditional costume.

co-operative. The sweating horses are led away. Then the branding begins. First the colts and fillies are separated from the mares and driven into a smaller yard. Then, while the mares neigh frantically, the iron with the letter K (for Kisháza) is pressed on their haunches sending out an acrid smell of burnt hide. One by one, they are allowed to rejoin their anxious mothers until at last the ordeal is over.

As the sun sets in a red glow on the distant horizon, the tired men and their wives go into the village in a small lorry and sit drinking local beer and eating a peppery stew with coarse white bread. They listen to the violin music of a gypsy band, and sing familiar folk songs almost as old as the plains.

11

Budapest and Other Cities

Excavations have shown that both banks of the Danube River where Budapest now stands were first settled in ancient times. Bronze Age and Iron Age man lived there and so did the Celts. The Romans built Aquincum (whose ruins can still be seen on the

A narrow cobbled street in Óbuda (Old Buda).

edge of Budapest). Aquincum was a busy trading-center in the Roman province of Pannonia.

After the Tartars invaded Hungary from the northeast in 1241, Buda on the hilly right bank of the Danube became a royal residence. In the second half of the 13th century King Béla IV built the stone fortress known as Buda Castle, in order to protect the city from further Tartar invasions. From then on, Buda grew in size and importance. By the 15th century it was a busy trading-center where merchants from east and west met. In the second half of that century, King Matthias Corvinus summoned artists and scholars to his court and turned Buda into a seat of learning. And when the Turks invaded Hungary, they recognized the strategic importance of Buda and so installed their pasha, or governor, there in 1541.

Buda and Pest were destroyed in 1838 by terrible floods when the Danube became blocked by an ice plug. After the floods, both cities were rebuilt and embankments were raised along the Danube. Pest, which lies on the flat left bank of the Danube, developed as a commercial and trading-center during the 18th century. This development was further stimulated by the coming of the railway age in the 19th century.

Already in the revolutionary days of 1848, Pest had been declared Hungary's capital. But it was only in 1873 that Buda, Óbuda (Old Buda), and Pest were joined together as a single city and proclaimed capital of the Kingdom of Hungary. Perhaps the greatest tragedy in the history of Budapest took place during the Second World War. Soviet troops and tanks captured Budapest

A view over the Danube, showing two of the six bridges which link Buda and Pest. The bridges were blown up in 1944, but rebuilt shortly afterwards.

from the Germans in the winter of 1944. After fierce fighting, most of the town lay in ruins and all the bridges across the Danube had been blown up. Over a third of the population fled to the surrounding countryside. Yet, soon after the war, in a few years, the whole of Budapest was rebuilt and all the bridges repaired and back in use.

Budapest again witnessed street-fighting, and tank battles between Hungarian and Soviet troops, in the autumn of 1956. Several buildings were destroyed by bombs and shell fire, but otherwise the damage was not great.

Today, Budapest has a population of 2,064,000 and is the largest city in Central Europe. It is not only a place of historical interest but also the seat of the Hungarian government. The president has his residence in Budapest and the prime minister and other ministers have their offices there. The Hungarian National Assembly holds its meetings in the Parliament Building in Pest, overlooking the Danube. The Royal Palace on the opposite bank, in Buda, has been turned into an art gallery and museum.

Buda and Pest are linked by six bridges. The famous Chain Bridge with its stone lions was built by the Scottish engineer Adam Clark. Elizabeth Bridge is a single-span suspension bridge. Árpád Bridge and Margaret Bridge are linked to Margaret Island. Margaret Island, known locally as *Margit Sziget*, has a park, sports grounds and facilities, two hotels, and a swimming pool fed by a hot spring.

In addition to the six bridges, Budapest has two railway bridges, an underground tunnel, and an underground railway. Budapest has wide roads running along the Danube embankments, as well as boulevards along which traffic flows past glittering superstores and great business houses. Trams and trolley-buses mingle with long lines of cars and trucks. The highest point in Buda is John's Hill which stands 1,736 feet (529 meters) above sea-level. The Liberation Monument is on Gellert Hill which is 771 feet (235 meters) high. Built after the Second World War, the Liberation Monument consists of a tall pillar surmounted by the statue of a Soviet soldier holding his rifle above his head.

The Millennium Monument is on the Pest side. This monument was built on the occasion of the 1,000th anniversary of the original settlement of Magyars in Hungary. It, too, consists of a tall pil-

The Millenium Monument, built to commemorate the 1,000th anniversary of the original settlement of Magyars in Hungary.

lar, this time with a winged statue on its top. Round it are statues of riders, horses, and chariots.

Pest also has the large City Park where there is a fun-fair and zoo. The fun-fair has a big wheel which is similar to the famous one at the fun-fair in Vienna.

Budapest has six universities and an Academy of Music. The city also boasts many museums which preserve the relics of Hungarian history. Among them are the National Museum (which has the Crown and Coronation Jewels on display), a Military History Museum, and a Catering Museum. Two other notable buildings are the Matyas Church, famed for the beautiful patterns of colored tiles on its roof, and the cathedral.

Several festivals are held in Budapest each year including the Spring Festival and the World Music Festival.

Steaming hot water, rich in radium and other health-giving minerals, rises from over one hundred thermal springs all over Budapest. As a result, people suffering from rheumatism and other illnesses come to Budapest to rest and take the baths prescribed by doctors. The Rudas Baths were once old Turkish Baths. The city's hotels provide luxurious accommodation for guests.

Because it has easy road and rail links with other parts of Hungary, Budapest has an excellent market where the farms can sell their vegetables and fruit. In exchange for their produce, the farms receive locally made consumer goods and agricultural machinery. There are flour mills and breweries in Budapest, as well as printing-works, and factories which make buses, precision instruments, textiles, and shoes.

About 112 miles (180 kilometers) eastward on the main road from Budapest, along the northern fringes of the Great Plain and then turning northward, is the market town of Miskolc. It is the third largest town in Hungary with a population of 200,000. Miskolc lies at the foot of the Bükk Mountains, near vineyards and a number of coal-pits. It is a busy road and railway junction. The land to the east and south is flat and parched in summer. However, there are oak and beech forests around the town and a spa with hot springs nearby. Apartment blocks were built in the suburbs to house the workers who came from the countryside to work in the new factories.

Some 142 miles (229 kilometers) from Budapest, also in eastern Hungary, is Debrecen, Hungary's second largest town. It has a population of 250,000. In 1848, Lajos Kossuth made the declaration of Hungarian independence in the great Calvinist Church in Debrecen. Behind the church is the Calvinist College, founded in 1546. The town is a mixture of old and new buildings with broad courtyards and gardens. The streets are wide and straight. Debrecen has a school of agriculture and a medical school. It holds a carnival of flowers every year in August.

The main road southward from Debrecen is straight and bordered by leafy trees. In summer, the countryside is green and yellow with ripening fields of wheat and maize. The villages have no

A horse-drawn cart—still a common means of transportation in Hungary.

squares or greens. They consist of whitewashed cottages with green shutters which line the road. Everywhere there are hens and chicks–on the road and in the ditches. There is not much motor traffic; the most common vehicles are horse-drawn carts and big combine harvesters. The road leads to the ancient town of Szeged with its cobbled streets.

Szeged is Hungary's fourth largest city with a population of 178,800 and it lies on the banks of the Tisza River, near the point where it is joined by the Maros tributary. Big floods devastated Szeged in 1879. Since then embankments have been built to protect the city from further flooding. There are smart shops and cafés in the center of this city which is famous for its medieval archway and churches. Being in the middle of a rich agricultural area, Szeged has many merchants' houses and a big cattle market. It also has a university where an open-air festival is held during the summer months.

Pécs is another town in southern Hungary, some 119 miles (198 kilometers) from Budapest and not far from the Croatian border. Every July a Summer Festival is held in Pécs. It was a known settlement at the time of the Romans who enjoyed its baths and hot springs. Today, Pécs is a busy industrial and market town with a population of 173,400. The countryside all around has neat fields in which red peppers and watermelons are grown. There are also fields of sunflowers and pruned rows of grapevines and fruit trees. Much of this produce finds its way to the market stalls of Pécs.

A general view of the town of Pécs in southern Hungary.

The town is proud of its university. Two of the churches in Pécs were mosques at the time when Hungary was ruled by the Muslim Turks.

Szolnok lies southeast of Budapest at a distance of 57 miles (92 kilometers). The town is built on the banks of the Tisza River and in Roman times it was a crossing-point between the provinces of Pannonia and Dacia. Today, Szolnok is an important railway junction with big marshaling yards. The town and its factories were badly damaged during the Second World War but all the ruined buildings have since been rebuilt. The town has been expanded with the establishment of new housing estates to house workers employed at the flour and sugar mills. There is a large swimming

A view of Györ on the Mosoni Danube.

pool at Szolnok for athletic competitions, and thermal baths which use water from underground hot springs. Szolnok also has a museum of folk costumes and an Aeronautical Engineering College.

Györ is an important industrial and cultural center which lies in northwest Hungary, on the main Vienna-Budapest route 79 miles (127 kilometers) from the Hungarian capital. Györ has a population of 129,600. It has a fortress at the point where the Rába River joins the Mosoni Danube, as well as a beautiful cathedral and palace. There are wide boulevards lined with trees, narrow cobbled streets, and a busy shopping center. The cafés and tea-shops have tables and chairs under colored sun-umbrellas on the pavement, surrounded by wooden troughs full of flowers.

12

Schools and Sports

As recently as 60 years ago, many Hungarian peasants over the age of 14 were illiterate. Hardly any of them went to school for more than two years. The moment they were able to look after themselves, peasant children were sent by their parents to work in the fields or to be herdsmen. So they grew up unable to read or write.

Boys and girls in both towns and villages attend kindergartens between the ages of three and six. Attendance at these schools is not compulsory but the great majority of children are sent to them by their parents. Compulsory education for the Hungarian child begins at the age of six. Basic or primary schooling, comprising general subjects with some practical training, continues until the child is 14 after which secondary education begins. Children must stay at school until the age of 16, but the majority of them continue their education beyond that age.

The most popular secondary schools are the grammar school, which is called *gimnazium*, and the vocational school, known as the *technikum*. The *gimnazium* has four classes in which academic sub-

A sailing regatta on Lake Balaton.

jects are studied. The *technikum* provides a general education but emphasis is laid on practical work. Factories and some farms have apprentice training-schools attached to them.

Hungary has a large number of higher education schools with 18 universities, including three technical universities, six agricultural universities, and four medical universities. There are also five university-level schools devoted to the arts.

Students in Hungarian schools learn much the same subjects as boys and girls elsewhere. They study the German language from an early age because of the traditions set by the Austro-Hungarian Empire in the last century. Emphasis is laid on scientific subjects.

Young people also have to study the Hungarian constitution. Constitution Day, on August 20, and October Revolution Day, on October 23, are observed as public holidays.

Schools in Hungary close in mid-June for the summer holidays which last until September. During the long summer months boys and girls go to camps on the shores of Lake Balaton or in the mountains. During the long dry summers they can swim in Hungary's lakes and rivers. Swimming and diving are popular in Hungary and many towns and holiday resorts have swimming pools, some of which are filled with the thermal waters found in many parts of Hungary.

Soccer is a popular game in Hungary and the Hungarian soccer team is greatly admired for its good record in international games. Because so much of Hungary is flat, cycling is another popular sport. Cycle competitions are organized between various clubs. Ice-skating is popular with both young and old people during the

Skating, sliding, and sledding on frozen lake Balaton.

cold winters and those living in towns can skate on specially prepared ice rinks on the local sports grounds.

Hungarians often hunt in the forests and in the marshes along the Danube and Tisza Rivers where there are flocks of wild duck, geese, and snipe. Shooting parties also go into the plains to hunt hares, pheasants, and partridges. Hikers can walk in nature reserves or national parks in the mountains.

Horse riding is not very popular in Hungary but long ponytreks are organized by riding schools. Hungarian sportsmen, however, are skilled at driving carriages and coaches. Hungary's top drivers won the World Driving Championships held near Budapest in 1984 taking the gold medal.

Every year, a medieval jousting tournament is staged at the Nagyvazsony Castle, a short distance from Lake Balaton.

13

Artists and Scientists

Music has a long tradition in Hungary. Folk music and dances are a feature of almost every festival and celebration. During the Middle Ages, princes, dukes, and nobility invited musicians to

Folk dancing at a local Hungarian festival.

their courts and encouraged them not only to perform but also to compose new music.

The Hungarian pianist and composer, Franz Liszt, greatly influenced younger musicians, such as Brahms and Grieg. Franz Liszt was born at Raiding, in 1811, and was taught to play the piano by his father. He gave his first public performance at the age of nine, and went to Vienna to study music. He toured France and, in 1848, was employed as Music Director to the Duke of Weimar. Liszt composed symphonies, piano concertos and his famous *Hungarian Rhapsodies* which were based on gypsy music. He conducted his oratorio, *St. Elizabeth*, in person at a concert given in Budapest in 1865. He was a great admirer of Beethoven whose bust he had on his piano. Works by Beethoven, Schubert, and Wagner were arranged for piano by Liszt. Finally, he retired to Rome and became a priest. He died at Bayreuth in 1886.

Another great Hungarian composer was Béla Bartók who was born in Transylvania in 1881. Bartók studied piano and composition at the Budapest School of Music. He was very interested in Hungarian folk music and, together with a fellow musician Zoltán Kodály, made recordings of folk tunes. Bartók composed piano and violin concertos, orchestral suites, and many piano pieces which were influenced by folk tunes. His *Third Piano Concerto* is popular all over the world. Bartók emigrated to America in 1940 and died there in 1945.

Another famous Hungarian musician was the violinist Joseph Szigeti. He was born in Budapest in 1892 and began playing the violin at concerts when he was only 13. He wrote a book about himself which he entitled *With Strings Attached.*

One of the greatest Hungarian painters of the 19th century was Mihály Munkácsy who was born in Munkacs in 1844. Munkácsy became famous for his large canvases. His best-known work is on a religious theme and is called *Christ before Pilate.* Munkácsy died in 1900.

Modern Hungarian artists make pottery and china of all kinds, including bowls and plates with complicated pattern painted on them. Woodcarvers make souvenirs for tourists as well as musical instruments. Many Hungarian women are skilled in making lace for blouses and babies' bonnets.

Ever since the industrial and scientific revolutions, Hungarian scientists have carried out research on medicines and diseases, on aerodynamics and nuclear physics. Robert Bárány, who was born in Vienna in 1876, was awarded the Nobel Prize in 1914 for his research work on the human ear. Another Hungarian Nobel Prize-winner was Albert von Szent-Györgyi who was born in Budapest in 1893. Szent-Györgyi was a biologist and professor at Szeged University. He won the 1937 Nobel Prize for chemistry for discovering Vitamin C in Hungarian powdered red pepper–paprika.

Hungarian nuclear physicists continue to carry out research on radioactive isotopes used in medicine and industry. Top scientists belong to the Hungarian Academy of Sciences with its headquarters in Budapest. Hungary is a member of the International Atomic Energy Agency which has its headquarters in Vienna.

Bertalan Farkas is an explorer and scientist. He was born at Gyulaháza, in northeast Hungary, in 1949. His father was a shoe-

The ornate interior of the Budapest Opera House.

maker. In his youth, Farkas took up glider-flying as a hobby. He was admitted to the György Killian Aeronautical Engineering College in Szolnok, and he graduated from there in 1969. He joined the Hungarian Air Force where he became a fighter pilot and was later promoted to squadron leader.

Farkas had very quick reactions and a high intelligence. On account of these qualities he was chosen to train as an astronaut in the Soviet Union. After months of training, Farkas and his Soviet fellow-astronaut Valery Kubasov were launched into space with the help of a rocket on May 26, 1980. They docked with the orbiting space-station *Salyut 6*. There Farkas carried out research on medicines and welding under weightless conditions. He also took photographs of Lake Balaton and stretches of the Danube and Tisza rivers to study the effects of industrial pollution. The two astronauts returned to earth a week later.

On his return, Farkas was given the title of Hero of the Soviet Union and Hero of the Hungarian People's Republic. Several other medals were also awarded to him.

The astronaut Bertalan Farkas and his Soviet colleague Valery Kubasov.

14

Rights and Duties

A new Hungarian state came into being after the Second World War. It was called the Hungarian People's Republic. The Communists took over the government with the help of the Soviet army stationed in the country. They banned all other political parties. The Hungarian parliament used to be called the *Diet*. Today, it is known as the National Assembly. This assembly adopted a new constitution on August 18, 1949 which was amended in April 1972.

In October 1989 Hungary became a republic and the constitution was again revised to reflect this new status. The new constitution proclaimed the economy to be a free-market with public and private property viewed as equal under the law. It also guaranteed the right of private property. The new constitution encourages labor unions to protect and represent the interests of the workers and entrepreneurs alike.

The constitution also states that the National Assembly decides how the country is to be governed. The National Assembly passes laws, approves the state budget and the national economic plan. The Assembly also decides when war shall be declared and peace made. So the Assembly is regarded as the highest form of state authority. It embodies the sovereign rights of the nation. The Assembly is elected for a term of four years and the members cannot be arrested or prosecuted without parliament's agreement. The sessions of the National Assembly are held in public.

The National Assembly (parliament) elects the president of the republic and the prime minister. The president appoints ambassadors abroad and concludes international treaties after their ratification by parliament. The president also can call for general elections. The president is the commander-in-chief of the armed forces, makes recommendations for the nomination of the president of the Supreme Court, and appoints judges.

The Hungarian constitution guarantees for its citizens the right to work and receive wages, the right to leisure time, the right to care in old age and in sickness, the right to schooling and equality before the law. Women enjoy equal rights with men. There is no discrimination on grounds of sex, religion, or race. In fact, racial discrimination is a punishable offense. The state is supposed to ensure the freedom of its citizens to go to church, to enjoy free speech and a free press.

Citizens from foreign countries who flee to Hungary because of political persecution enjoy the right of asylum.

All Hungarian citizens have the duty to protect public property and work towards raising the standard of living of the people. They also have the duty to do military service and defend their country in time of war.

The Parliament Building in Pest, in which the Hungarian National Assembly holds its meetings.

The army has nearly 100,000 officers and men. The private soldier is called *honvéd* which means "home defender." The air force has 12,000 men but Hungary has no real navy, only a few river gunboats.

Modern Hungary has a national insurance scheme. There are social services for old people, and for the disabled and victims of disasters such as air crashes, train accidents, floods, and earthquakes. A uniform social security system for all was introduced in 1975.

Hungary is a member of the United Nations Organization and has always taken an active part in the proceedings of the United Nations General Assembly. Until recently Hungarian representatives at the United Nations traditionally voted for resolutions put forward by the Soviet Union and other communist countries, with the exception of China. The modern Republic of Hungary is now independent of those restraints.

15

Hungary and the Modern World

Half of today's Hungarians are still peasants. Compared with workers living in the west, they have low incomes, and few cars and television sets. Their country is poor in natural resources, such as minerals and energy fuels. Hungary is a developing country in need of oil, coal, and electricity. Energy costs a lot of money in the modern world. Hungary must increase her exports of manufactured goods and farming produce to find the money to pay for energy. The government is trying to find new ways of increasing electricity supplies.

Developing new industries with modern equipment creates its own problems–it needs trained workers to operate computers and automatic machinery. So young people must attend new courses at technical schools at home and abroad to understand new technology.

Housing in towns is also a major problem in Hungary. The industrialization has drawn peasants from the countryside into towns and they need homes. New housing estates have been built but there is still a shortage of accommodation, of water supplies and other public services. Whole families sometimes live in two rooms.

Hungary has signed important trade contracts with Britain, France, Germany and other western countries and borrowed millions of dollars from western banks for economic development at home. Foreign firms have been encouraged to become partners with Hungarian firms in developing the tourist industry, farming, and modern technology. In the future, more capital will be put into engineering, electronics, telecommunications, and pharmaceuticals. This policy will make it possible to improve the standard of living of the Hungarian people.

More crops are being produced on farms and there are more goods in the shops today, though people still complain of shortages. There are free nurseries for children and free medical treatment for all. Hungarians can own their own businesses, and so become richer by working hard. More and more people in Hungary today have new cars and vans, though waiting lists are long. Some of the more fortunate Hungarians own a second house on the banks of the Danube or in the mountains.

The new republic is struggling into the 21st century trying to resolve its economic concerns, and inflation is beginning to drop. Hungary is strengthening its ties with the Western world. It has applied for membership in NATO and is planning on joining the European Union (EU).

Work on the countries infrastructure is a priority with the

upgrading and extention of the nations highways, airport expansion, and updating of their antiquated telephone system planned.

In the elections of 1998 the Hungarians put their trust in a new generation to take them into the next century–hopefully a century of growth and freedom.

The Hungarians are intensely patriotic. Having gone through many foreign invasions in the past, they now believe in being a sovereign nation and free people. Some feel that the world has been unjust to them because so much Hungarian territory was taken away from them after the two world wars. As a result, many Hungarians live outside their country's borders and are citizens of foreign states. Yet the government accepts that the present borders

Painted shops and houses—a style which can be found in Hungary and several other central European countries.

are final and makes no claim to territory from any of Hungary's neighbors.

The Hungarian people have preserved their faith in their nationhood and in their traditional way of life. They are a funloving people, fond of music and good living. They spend much time in cafés drinking black coffee and iced water and sampling rich chocolate cakes. Though they live in the midst of a dangerous and violent world, they feel confident that their future stability is assured.

GLOSSARY

Communism A totalitarian system of government that eliminates private property. It is based on Marxian socialism.

csikos Cowboys

Diet Hungarian parliament

Huns Mongolian people who invaded the plains of Europe in the fifth century. They conquered the Magyars and took over their land. The word Hungarian is derived from the word *Hun* but the Hungarians as a people are not descendants of the Huns.

hydrofoil A boat that travels by skimming the surface of the water

karst An irregular limestone area with underground streams and caverns

loess A yellowish-brown soil that is extremely fertile but easily blown about by the wind

Magyars Hungarians

mosque Place of worship in the Muslim religion

Puszta The Great Hungarian Plain that lies to the east of the Danube River. The word *puszta* in Hungarian means "waste land."

steppe Vast, usually level, treeless tracts of land found in southeastern Europe and Asia

INDEX

A
Adriatic Sea, 32, 41
Aggtelek Mountains, 20, 21
Agriculture, 22, 23, 26-27, 28, 51, 53, 59-66
Alföld (Great Hungarian Plain), 19
Alpine Mountains, 19
Andrew II, 35
Antal, József, 12
Aquincum, 33, 71-72
Arpád, 33, 34, 35
Asia, 35
Attila the Hun, 9, 31
Austria, Austrians, 19, 22, 39, 41, 42-43, 49
Austro-Hungarian Empire, 11, 43, 44

B
Bakony Forest, 15
Balaton, Lake, 14-15, 23, 55
Bálvány Peak, 20
Bárány, Robert, 87
Bartók, Béla, 86
Batu Khan, 35
Bauxite, 24
Béla IV, 35, 72
Bohemia, 35, 37
Bridges, 74
Britain, 44, 47
Buda, 37, 40, 72. *See also* Budapest
Buda Castle, 35, 72
Budapest, 27, 33, 49, 52, 71-76
Buffaloes, 23, 27
Bükk Mountains, 19, 20, 76
Bulgaria, 9, 33

C
Carpathian Mountains, 14, 19, 33, 41, 44
Cattle, 63
Charles IV, 45
Charles of Anjou, 35
Christianity, 9, 34
Climate, 24
Clothing, 16
Coal, 24
Communism, 11, 47-48
Corvinus I. *See* Hunyadi, Matthias
Croatia, 19, 27
Csepel, 26, 52
Csikos, 69

Czech Republic, 43
Czechoslavakia, 46, 49. *See also* Czech Republic and Slovakia

D
Dacia, Dacians, 32, 33
Danube River, 9, 11, 13, 19, 25-30, 32, 52, 55, 58, 60, 71
Debrecen, 53, 77
Dinaric Mountains, 19
Dráva River, 19
Dripstone Cave, 21
Dual Monarchy, 43

E
Economy, 15
Education, 81-83
Europe, 13, 19, 22, 26, 42
European Union, 95

F
Farkas, Bertalan, 49, 87-89
Farms, farming. *See* agriculture
Ferdinand I (Hapsburg), 39
Ferihegy, 55
Fertö, Lake, 22
Festivals, 85
Finns, Finland, 15
First World War, 44, 45
Fish, fishing. 58
Five-Year Plans, 51
Flag, 47-48
Forests, 56, 58
France, 44
Francis Ferdinand, Archduke (Austria), 11, 44
Francis Joseph, Emperor (Austria), 11, 43
Fruit, 62

G
Germans, Germany, 9, 11, 16, 26, 33, 44, 46-47
Great Hungarian Plain, 13, 14, 19, 21-22, 33
Györ, 54, 80
Gypsies, 15

H
Hapsburg Empire, 10, 41
Henry V (Germany), 33
Herbs, 62-63
Hitler, Adolf, 46
Horses, 64-65
Horthy, Nicolas, 46, 47
Hunedoara Castle, 37
Huns, 31
Hunyadi, János, 36, 37
Hunyadi, Matthias, 37, 72
Huss, Jan, 10, 35

I
Illyrians, 32
Industry, 51-54
Irrigation, 28, 60

J
John Sobieski III (Poland), 40

K
Kádár, János, 12, 49, 50
Kápolnapuszta, 23
Karlowitz (Serbia), 40
Károlyi, Mihály, 45
Kékes, Mount, 20

Khazars, 31
Kossovo, 37
Kossuth, Lajos, 10, 11, 42-43
Kun, Béla, 46

L
Lechfeld, 9, 33
Leopold I, Emperor (Austria), 40
Liberation Monument, 74
Liszt, Franz, 86
Little Plains, 22-23
Loess, 22
Louis, King, 37

M
Magyars, 9, 15, 31, 35, 74
Margaret Island, 74
Maria Theresa, Empress, 10, 42
Mátra Mountains, 19, 20
Maximilian II, Emperor, 10, 39
Mecsek Mountains, 23, 24, 63
Merseburg, 33
Migration, 33
Millenium Monument, 74-75
Mindszenty, Cardinal József, 12, 48
Minerals, mines, 15, 24, 37
Miskolc, 52, 76-77
Mongols, 10, 31, 35
Mosoni River, 26
Munkácsy, Mihály, 87

N
Nagy, Imre, 49
National Assembly, 47, 74, 90-91
National Council, 45
NATO, 47, 95
Northern Mountains, 19, 56, 63

O
Óbuda, 72
Oil, 24
Orbán, Viktor, 12
Otto I (Germany), 33
Ottoman Empire, 10

P
Paks, 27
Pannonia, 9, 23, 32, 33, 72
Pécs, 33, 53, 78-79
Pest, 37, 54, 72, 74, 75. *See also* Budapest
Petöfi, Sandor, 42
Pied Piper of Hamelin, 37
Poland, Polish, 10, 33
Puszta, 13, 19

R
Rába River, 26
Rakoczi, Prince Ferenc, 41
Railways, 54
Religion, 13, 34, 35
Revolution of 1956, 12
Roman Catholicism, 13, 25, 34, 48
Romania, Romanians, 16, 19, 44, 46
Romans, 9, 32, 33, 71
Russia, Russians, 15, 31, 43, 44

S
Sarajevo, 11, 44
Saxons, 37
Second World War, 16, 46-47, 72-73
Selim II (Sultan), 39
Serbia, 19, 27, 36, 37, 44
Sigismund of Luxembourg, 35
Slovakia, 18, 19, 20, 26, 43

Slovenia, 19
Soviet Union, 12, 16, 24, 46, 49, 73, 90. *See also* Russia
Space travel, 49
Sports, 83-84
Stephen I, 9, 34, 43
Szatmár, Treaty of, 41
Szeged, 60, 78
Szeklers, 36
Szent-Györgyi, Albert von, 87
Szmobathely, 33
Szigeti, Joseph, 86-87
Szolnok, 54, 79-80

T

Tartars, 35, 72
Timber, 57
Tisza River, 22, 27, 58, 60, 78, 79
Transdanubia, 15, 22, 23-24, 32
Transportation, 28-29, 51, 54-55
Transylvania, 10, 34, 36-37, 38, 39, 43, 46
Turkey, Turks, 10, 27, 34, 36, 37, 39-40, 72
Turkish Empire, 44

U

Ugrians, 31
Ukraine, 19, 22
United States, 47
Uranium, 24

V

Vác, 54
Vienna Arbitration Award, 46
Vineyards, 24, 63
Volga River, 15, 31

W

Warsaw Pact, 12, 47
Wildlife, 21

Y

Yugoslavia, 32, 43, 46, 49

Z

Zala, 24